Mini Corpus Venn 4

Contact Information:
diogodesouza7@gmail.com
diogodesouza7@hotmail.com

Introduction

This book is a short introduction to the Fluidal Nature of Reality, where Energy and Matter propagates in Space-Time to generate Particle Interactions like the flow of Fluids. It is the idea that both Light and Sound are Waves, and Waves imply Oscillations which may or not involve a Medium to propagate. These Disturbances in Space carry Energy, Frequency, and Pressure and that is the way by which Particles experience Forces. There are Particles that are Force Carriers such as Bosons, and there are Particles that experience the Forces which give form to all the Matter that we see. The Universe is a Grand Design where all things can be understood Mathematically and with the concept of Waves and considering all things as Cyclical like vibrations in strings that repeat over time. There is then a brief introduction to Quantum Mechanics, proof for Taylor's Series and Euler's Identity, and the usefulness of Imaginary Numbers to describe Waves. There is also comments on Standing Waves, Energy Levels, and Matrix Algebra such as finding Eigenvalues and Eigenvectors. This is a minuscule portion of an Universe that is really complex but that can be understood through reason and through a series of logical steps and that is the goal of Science.

One Essence

There is a Source of all things in the Cosmos which is the Essence that determines the character and function of every part of the Universe. All parts of the Universe comprise portions of the whole. In the same way that a living Organism is composed of parts, the Universe is in all of its complexities made of parts. In the same way that all the parts in a Living Organism depend on each other to maintain Homeostasis, the same can be said on all the Fields of Knowledge with subjects with contents that overlap. There is Physics in Biology, and Chemistry in Physics, and so forth since nothing that exists is in isolation. All things depend on all things. Everything is deeply connected. In Essence everything is derived from the same Ultimate Principle. Everything is Matter and Energy. From a Source of Light which is Energy there was the formation of Matter and the differentiation of the Whole into individual parts that together forms the whole now more complex with more defined parts, including differences, opposites, characteristics, and different purposes that work together in their support of everything.

Substance

The Essence of all things is a substance in the realm of forms that leads to the being and the natural Subatomic behavior found within everything. In the beginning there was Light, and Light Particles formed all the Matter in the Universe. Particles are vibrating strings of Light and these vibrations in Hilbert Space when added together generates the many objects, types of Matter, and lays the foundation for Chemical and Physical Properties with nature obeying the many Laws and Constants that sustain and moves the flow of the Cosmic Structure Machine. In this maze of Particles of Light and Chemical Reactions, the Space Time continuum can be found held in place through macroscopic sized Matter clumped together creating a reality for us to perceive the flow of Time and the movement through Space. The Universe is Mechanical and comprises several Dimensions, many of which are curled up at the Subatomic World with only four Dimensions visible and perceived daily. There are three Dimensions of Space and one of Time. In order to explain the great varieties of reactions of Particles and the flow of Energy at the Microscale, the understanding of many other Dimensions is necessary.

The reality is the projection of Light into the eyes and feelings of the observer. Behind this projection there is a complex code with instructions as to how Nature ought to behave. These instructions are Mathematical Equations and Geometry that leads to an Infinite summation in the forming of Fractals, shapes in a Sea of Fluctuations, and the origin of many Forms. Everything is composed of the same Substance called Light, but at different amounts, different Frequencies and Wavelength. Like several Radio Stations that have in common Light as their Source of Electromagnetic Communication. What makes each Particle and each soul unique is a different Vibrational Pattern. The groundwork for all Natural Phenomena is multiple Equations, and the projection of these Rules into our Reality is what we perceive as being alive. The Substance is the Essence of all things, and these are Waves of Energy that can be transformed into its Multiple Forms. There is a rich content of details in all of Geometry surrounding all causes and effects, weather, flow of blood, flux of Electric and Magnetic Fields. Wisdom is found in the Flow of objects, the flipping of pages in the grand theater held in place by the pillars of the Four Forces of Nature which are Gravity, Light, Weak, and Strong Force.

Wisdom is a Flux

Humans can only read if a flow of Light Waves enters our eyes, and if a flow of Chemical Reactions in our brain between Neurons leads to perception. Life can only exist if there is a flow of air containing the right ingredients and the right Temperature for life to flourish. All things are in Essence the flow of something. There are flows everywhere and these flows carry Density, Pressure, Mass, Energy, and Velocity. These flows lead to entanglement, interconnectedness, and the many forms of communications between living beings that exchange Particles. Even nonliving things communicate such as how Particles react to each other in such a great way that is capable of forming beautiful rings and spirals like the DNA Molecule. The existence of the Universe is forever, which is enough time for the formation of all things, and anything that is possible can and maybe will happen. Our hearts are constantly pumping Blood, which flows in Organisms exchanging Energy. There is a great diversity of things and all that exists shares the same embedded nature which is Light and Flow of Energy.

Fluids

When a person speaks or a dog barks, a vibration produced in their vocal cords leads to a disturbance in the air, and the flow of this disturbance leads to the propagation of Sound. Light is also a propagation through the vibrating Dark Matter in Space. Dark Matter is the Ether, and all things are made of this Ether when it compresses and becomes visible or detectable through instruments. Fluids are all around us and the roots of reality, which is what constructs our Space Time Dimension, is the flow of fluids, Light and Sound, Gravity and Electromagnetism, exchange of Particles, and the rotating of wheels. The Universe is a Machine, but the workings of this Machine is hidden from us. It is like living in a Virtual Reality and trying to learn about it, without realizing that you are with a helmet over your head and sitting in a white cold room. The Universe is just a projection, and our perceived reality is not its true workings. Unless we get to see the rotating wheel and the flow of charges through the circuits that allows the functioning of reality, we are totally blind to its true nature.

Electrodynamics Flux Divergence and Curl

The Forces between Particles are from the exchange of Light emanating and filling the Space with a Fluid of Energy. Since there are Particles everywhere and in all directions that we can look, we literally live in a Sea of these flows. Truly Wisdom is a flow, perception is a flow, reality is a flow, and there are also wheels rotating, and geometries obeying symmetries, and Infinities and Eternities overlapping each other, with entanglement, and interconnectedness. Magnetic and Electric Fields are Fluxes through Space Time, with the Magnets being Dipoles and the Electric Field also being capable of Monopoles. Every time we hear something new, a flow of Energy arrives in our Antennas, and is perceived by our minds, which once again is a Flow of Charges between Neurons. These lead to reactions which is a discharge of Energy in the Muscles. Discharges are also flows, and there are fluids everywhere. Information comes in waves, disturbances in the Quantum Field of Space where all history occurs. Reality is where all existence is located and prosperity comes with learning experience, new outcomes, which is the evolutionary process and transformation through time.

Vectors and Geometry

While living in a Sea of Quantum Fluctuations, discharges of Energy, transfer of information, Chemical Reactions, the eternal Flow of thoughts and Vibrational Patterns, Vectors then gives direction of a Field at a given instant. The collection of all Wave Functions is the summation of several Vectors. Since Wisdom is now acknowledged to be a Flux, and since any flowing fluid has a direction of propagation even if from a point source, then Wisdom is also a Vector or a collection of Vectors. There is both magnitude and direction for Forces, Fields, Light, and Sound. The Sea of Fluctuations forms a Basis from where all the possible Vectors in a particular Dimension can be formed. All Forms in the Universe are derived from a higher Substance. I said earlier that this Substance is Light, which is Energy, which is now a Flux and recently now is also considered a Vector. We can summarize then that Fluxes, Light, and Sound Propagation can be viewed as Vectors interacting with each other in the form of Wave Functions. When the intensity of this Energy is great it can be dangerous for living beings since it becomes highly radioactive. Energy and waves are beneficial in the right amounts. The extremes of anything are never good.

Living inside a Black Hole

Albert Einstein was the greatest Genius of the 20th Century, and his Theory of Relativity has profound implications. Just like any Theory, it can be expanded and later become an inspiration for multiple other ideas in the future on how the Nature works. In the same way that there are millions of books inspired by Aristotle, the Theory of Relativity is a light at the end of a tunnel for Space Travel, placing Satellites in orbit of Planets, and an infinite number of thought experiments. Knowledge should grow after each generation, and a Library is a collection of Human Critical Thinking through Time. Even if the book is a fiction, it is a thought experiment, a comic book is a thought experiment, a game is a thought experiment. So, we can clearly say that all reality can also be viewed as a thought experiment. Hopefully the experience is a peaceful one. Wisdom besides being a flux, it is also a thought experiment. According to the Theory of Relativity, in each Planet, in each Dimension we have a unique perception of the Flow of Time and the rate of the motion of objects through Space. In other words, we can say that in the Universe there are many individual worlds, each with its own version of reality, thus the roots of Relativity.

If we take in consideration that everything that a Black Hole swallows a White Hole spit out, then the Big Bang was the spitting of Matter and Energy from a White Hole, and in the future, we may be swallowed by a Black Hole or even more, we may be living inside a Wormhole right now. The Universe and everything in it, is like a Bubble travelling through Space and Time inside a Wormhole heading towards the Big Crunch future. Then we can conclude that there is no Universe without a Black Hole, and knowing that time stops inside a Black Hole, which is in perfect agreement with the fact that Time starts at Time zero, and after that, the Time begins flowing. What if then the Universe is like a Quantum Loop? Is it just pure coincidence that the Symbol for Infinity is a Loop? Human perception of Reality whether Ancient or Modern, holds secrets that only later humanity acknowledges. These are beliefs later found to be in fact true. The Universe, however, is a whole lot more creative than what we can ever be. We can be inspired by the Cosmos and learn from it to gain insights into Reality, many of which in fact True and only Time then clarifies what it means with a shining light at the end of the Tunnel. ∞

Everything is Art

In the same way that what makes Art beautiful are the richness of details, and the evidence for a significant amount of time used to generate it, the Universe and all of Wisdom is also a great work Art. Appreciating the Art is akin to contemplating a portion of the Cosmos. When we look at a mountain, we contemplate what it means, how it was formed, or when did it gain its shape. The great Philosopher called Socrates once said that a life without observation is not worth living. Contemplating a work of Art is also the same as performing the Art. There is no Art without contemplation. The goal of all traces, colors, shapes, is to represent or transmit a feeling to the observer. There are forms of Art that can ignite feelings of peace, others of joy, others of inspiration, and similar to music Art is a way to communicate feelings. Similar to how reading is a way to absorb information, and to have a thought experiment, Art works sends visual content, and Music are information in the form of Sound. When reading we can hear a voice in our head that reads, and we may, in our imagination, see the event that the book describes. Life in the like manner is a thought experiment. All things are Art and its contemplation.

Axioms and Definitions

In the beginning everything was the same one thing, the Substance at the Singularity, a Particle Soup in the form of Energy. With the expansion of the Universe, the one thing began to differentiate and to give form to multiple things leading to distinction, hierarchies, and different Elements, Atoms, Matter, and Energy of a variety of Frequencies. Each thing, each Element is something and is not something else. A man is a man, a dog is a dog, a man is not a dog, and a dog is not a man. All things are something and are not something else. In order to define the characteristics and nature of specific things, it is necessary the usage of logic, labels, statements, to describe the diversity and complexities that makes the Cosmos the grand Theater of Forms, and Events. Through logical steps we can arrive at conclusions that can be general or that can be more specific. Not all men are tall, but some are tall. That is a specific statement. Now, all men are Homo Sapiens. That is a General Statement which is correct and that is a broad definition. General Statements can also be wrong when conclusions are taken without being careful to other possibilities.

For example, if bob is a man and has black hair, then since Joseph is also a man, he must also have black hair. That is an example in which a General Statement is wrong for not being careful to many other possibilities such as also being a man but with red hair. General Statements include everything and are at a greater risk of being wrong. When defining something more specific, claims are at a lesser risk of being wrong, but even then, there is a chance that observers can claim a conclusion that misses the truth by a given amount. The Ancient Greek Sophists believed that man is the Measure of all things. We humans place labels in all things, we define things, we distinguish, we try to understand, we are constantly learning from observations, and there are Measurements even in our actions, and our moods are Measurements too. How happy a person is, how smart an individual is based on his or her IQ. Also, we Measure when analyzing what to do, what is the best thing to say, or who to spend time and talk about something. All of these everyday actions and reactions to life events is a form of Measurement. When we Measure, we live, and we can't live without Measurement. Even a baby cries when hungry knowing that a Measurement is needed to fill the stomach with more milk. We are always Measuring and defining, seeking, and evolving through Time.

Space

Space is a great void comprised of innumerable locations where this void is compressed generating shinning lights dotted in every direction. Anything that is present in Space emits light. Stars shine while Planets, Asteroids, and Comets emit light that they absorb from the Sun. All objects emit light, humans for example, emit light that peak at Infrared Radiation which is heat. Particles emit light in order to generate the Electric and Magnetic Fields surrounding them. The emission of Light is what leads to the Four Forces of Nature. Even Gravity is a form of Light propagation with Waves that bend Space, constructing the several distinct Space-Time Dimensions. Gravity bends by generating waves and ripples along the Strings of Space which are the same void and Strings that are used by the Photons of Light to give form to the Fields that leads to the other Forces of Nature. Since Particles are constantly interacting with each other, and since all Matter is made of Particles, then all Matter emits light. Stars shine but many other objects do not shine at the Visible Light Spectrum on their own but still emit Light since they all have heat.

Phase Space

Everything in the Universe undergoes Oscillations. The vibrating strings of Particles, the orbit of Planets around Stars, the pumping of blood in Organisms, and the classical example of Pendulums and Springs are few of the multiple forms of vibration and Cyclical Patterns seen all around us. Even the Universe could be cyclical with birth, growth, death and birth again. All of the Natural World is quantized. The Phase Space of a Simple Harmonic Motion when graphed with Momentum in the vertical axis and speed in the horizontal axis has the form of an oval or a circle whose area is quantized. There are different modes of Vibrations and each with a unique line plotted on the graph. The different steps of Oscillations starting with the Ground State with the least Energy and going up in steps to higher states, each state with a specific pattern of a line on the graph. The difference in areas of any two different states are multiples of the Planck's Constant. That means that all vibrations, all laws of the Universe are quantized existing in discrete steps. The Universe in comprised of bits of information, degrees of Energy, and it appears smooth since the steps, the bits, are very small.

The Phase Space graph of Energy versus Time can't be less than the Planck's Constant. The Phase Space graph of Momentum versus Position also can't be less than the Planck's Constant. The smallest value which is the Planck's Constant represents the Ground State of the Oscillation which is the smallest Energy a System can have above Zero Energy. The difference in Energy of any two states above the Ground State are multiples of the Planck's Constant. This discreteness exists all around us, and these are the pixels of the forms and characteristics found in all Natural Phenomena that is perceived. Phase Space is a human attempt to label, predict, and explain the Quantum reality. Quantum Mechanics is a tool with Theories and Equations that provides us with approximations to understand the Dimensions of Space and the behavior of the Subatomic World. The Universe at the very small scale is a Sea of Fluctuations that can only be understood with Quantum Field Theory. At the Macro Scale, the Theory of Relativity explains Gravity and the large structure of the Cosmos. The unification of both Theories provides Scientists with a Theory of Everything. Where does Gravity and Quantum Mechanics overlap? They overlap possibly within Black Holes, under intense Gravity, at a very small size such as in a Singularity.

Energy(Time) = Momentum(Velocity) = Planck's Constant at the Ground State.

In a Simple Harmonic Oscillator, the multiplication of Energy times the time is equal to the Momentum times Velocity which at the Ground State is equal to the Planck's Constant. At higher Energies, the Vibration or Oscillation Pattern exists at increments of Planck's Constant such as 2h, 3h, 4h, and so forth towards Infinity with h being the Planck's Constant and the number in front represents the Energy Level. The fact that the Natural World is generated through increments, is evidence of the Discreteness which is the only reason why it is within human reach to understand the structure of the Cosmos. If nothing were discrete mathematics would be worthless since the lack of discreteness would mean that it would not be possible to measure things or to design a geometrical framework to learn about reality. It is the fact that things exist in increments that allows us to use mathematical equations and make predictions even within probabilities. Despite of the fact that Quantum Mechanics teaches that nothing can be 100% certain, we can still be certain at what the probabilities are, however.

Closed System

Closed Systems are ones in which nothing from outside is allowed in, and nothing from inside is allowed out the System. Imagine a box enclosed in such a way that is a perfect barrier against external forces and is sealed against the exchange of Particles through the barriers. Suppose that inside this box there is a Sea of Particles. The motion of these Particles can be represented with Vectors with both Magnitude and Direction. The Direction of the Particles is the same as the direction of their Velocities and they also carry Momentum. Assuming a Closed System and that all collisions of Particles with other Particles and their collision with the walls of the box are Elastic, then no Energy is lost from the collisions and the System is in perfect equilibrium. That means that Energy, Momentum, and the sum of the Forces are all conserved. Even the Particles moving randomly and colliding, bouncing off each other, makes the sum of all the Vectors of Force, Momentum, and Velocities equal to a constant value. In real life a perfect Isolated and Closed System does not exist since Energy is always lost due to Friction, Heat, and Sound, but what if the Universe is Closed System?

Nothing within the Universe is a Closed System. A box will always exchange some form of heat with the environment outside of it. What technology can do is create a seal that will reduce the exchange of Heat and Matter significantly, but in the same way that the Universe only moves to a state of higher Entropy or disorder, the Natural Law is for there to be impossible to create a machine 100% efficient or a box that is 100% sealed. There is always Matter and Energy leaking between barriers. When stating the same for the entire Universe is where there can be controversy. If there are no other Universes besides our own, then it makes sense to conclude that since everything is found within a Universe with no leakage of Matter and Energy to and from its outside, then the Universe is perfectly closed. The case however can be flawed with the belief of the Multiverse. If there are multiple Universes, then a leak of Matter and Energy between Universes could in fact happen. Quantum Mechanics states that the Superposition of Particle Quantum States is an indication that these Particles exist in multiple Universes at the same time, and being these states entangled, the Universes communicate with each other, and no Universe is perfectly closed.

Assumptions

The Universe is more complicated than what can be explained by Theories devised by Scientists to better understand reality. When attempting to synthesize and measure things in the cosmos, Scientists may assume Ideal Cases. When studying the relationship of Volume, Pressure, and Temperature of Gases, Scientists may first assume a simpler model. These Ideal Cases are never the truth since there is no such gas in the real world. The Ideal Conditions ignore Energy losses, and no conservation of Angular and Linear Momentum, Velocity, and Forces. When studying the function of a Pulley and how it can be useful to explain Net Force and Acceleration, Scientists may once again assume an Ideal Situation such as in the example of the Atwood Machine. No real Pulley obeys the Atwood Machine. This example ignores Moment of Inertia, Mass of String, and Friction. Despite of the fact that Ideal Cases are never the truth, they are a good start into understanding more complicated aspects of Reality. We can only understand the difficult things if first we are able to grasp more basic and Ideal Conditions. It is common then for Scientists to make assumptions.

Probability Clouds

Particles can be represented with Waves. Oscillations can be plotted in a graph using coordinates in two or three Dimensions. To eliminate negative Amplitudes, Wave Functions can be squared. The Square of these Waves gives you Probabilities, and a Constant placed in front of this Probability Cloud is to allow the chances of finding the Particle somewhere in the Universe to add up to exactly 100% Percent. There are the Quantum States, and when an Operator acts on the Wave Function it leads to a specific Eigenvalues whose Absolute Value Squared leads to the Probability that the Particle will be in that state. Wave Functions are a collection of other Wave Functions each representing a Quantum State among a Sea of Fluctuations and Possibilities. The smallest number between two numbers is their difference. The Uncertainty Principle defines that it is not possible to absolutely know both the Momentum and Position of a Particle based on its Wave Function that occupies a blurry region of Space Time. At the Subatomic World, Quantum Effects are evident since the Planck's Constant is small and that is where Uncertainties exist in the Sea of Fluctuations.

Matrices

We live in a Matrix. The Dimensions of Space and Time can be mapped with Vector Components within Matrices. The Quantum Sea of Particles and Forces is described by Matrices. The Basis Vectors of a region in Space, and the collection of Particle Quantum States, all can be understood as Matrices. The Computer Simulations runs a collection of 0s and 1s, a Binary Code within a Matrix. All of the Universe is comprised of series of Bits of Information, similar to a Computer Simulation. We are inside a huge Matrix controlling and holding in place all the Forces of Nature and all of Natural Phenomena. This is the realm of Trigonometry, and the construction of Reality depends on the Space and Time Geometry. Mathematics is everywhere running our lives, our existence, and the order of events obeying a determined order from Equations and Constants. These Equations and Constants are the signature that makes existence a real experience. There are Laws that Particles follow in their interactions with other Particles. In the same way that we humans obey orders in our human world, Nature is not at random but orderly.

Linear Algebra and the Grid

The Universe is a huge Matrix, like a Computer Software. A Matrix is like a Grid with Rows and Columns containing Information. When reading a book we read letter by letter, word for word, line by line. Books contain Literature in the form of a Grid, with Rows and Columns with lots of Information. Everything that is and exists is the result of a collection of codes in the Grid. We all live in the Grid that is so perfect that it is possible to live entire lives without noticing it. All knowledge is a compilation of Bits of Information in a Grid. Our bodies are Organisms with Cells, Molecules, and Atoms in a Grid Pattern. Since we are all in a Grid, we are all connected to each other. The Study of Matrices and Grid relies upon the Science of Linear Algebra which I am about to introduce in this book. Linear Algebra provides the connection between Math, Geometry, and Trigonometry. The Science of Matrices is the key to everything in the Cosmos. All things well measured and calculated. It is the Logos Cosmos everywhere we look, and everything we feel. Emotions are experience withing the Grid of our Brain Neurons. We are Matrices as well.

Linear Algebra Part 1

Matrices are a set of Vectors that are lined up according to their components. They are comprised of information, Ratios, and Proportions and one whole Matrix can be a linear combination of other Matrices.

Suppose we have Vector **A** = ax + by + cz, Vector **B** = dx + ey + fz, and Vector **C** = gx + hy + iz

If we place them in order, we get:

$$\begin{bmatrix} ax & by & cz \\ dx & ey & fz \\ gx & hy & iz \end{bmatrix} = \text{A } \textbf{Matrix} \text{ with the combination of}$$

these Vectors. It can also be written in the form:

$$\begin{bmatrix} a & b & c \\ d & e & f \\ g & h & i \end{bmatrix} \begin{bmatrix} x \\ y \\ z \end{bmatrix} = \textbf{Matrix} \quad \text{Which can be described as a}$$

Linear Combination:

$$x \begin{bmatrix} a \\ d \\ g \end{bmatrix} + y \begin{bmatrix} b \\ e \\ h \end{bmatrix} + z \begin{bmatrix} c \\ f \\ i \end{bmatrix} = \textbf{Matrix}$$

Here we have three Vectors with three Dimensions x, y, and z that comprise a single Matrix.

Solutions of a Matrix:

The Vectors in a Matrix are part of a line of codes. These lines are composed of Rows and Columns. The Columns could indicate the Dimension such as x, y, or z, and the Rows could represent the Vector such as Vectors **A, B**, and **C**. The relationship between Vectors could be Ratios or Proportions, or Vectors could have no relation to other Vectors inside the Matrix. When Vectors are not a combination, or a Ratio of the other Vectors, the Matrix is called Linearly Independent. When Vectors are a combination or a Ratio of the other Vectors, the Matrix is Linearly Dependent. If there are three or more Vectors in a Matrix, and if at least two of these Vectors are a Linear Combination of the other, then the whole Matrix is Linearly Dependent. If there are two Vectors and they are a combination of each other, then the Matrix is also Linearly Dependent. Linearly Independent Matrices only have the Trivial Solution when the Matrix is set equal to zero. Linearly Dependent Matrices can have an Infinite Number of Solutions and Matrices could also have no Solution.

Linear Independent Matrix:

Suppose that there are three Vectors that are not a Linear Combination of the others.

$$\begin{bmatrix} 1 & 2 & 3 \\ 7 & 8 & 9 \\ 3 & 7 & 9 \end{bmatrix} = \textbf{Matrix}$$

Using ratios, multiplication, division, and addition, the Matrix above is simplified in the form:

$$\begin{bmatrix} 1 & 0 & 0 \\ 0 & 1 & 0 \\ 0 & 0 & 1 \end{bmatrix} = \textbf{Matrix}$$

Which means that if the Matrix is set equal to zero only the

Trivial Solution = $\begin{bmatrix} 0 \\ 0 \\ 0 \end{bmatrix}$ exists.

The Matrix on the right Span R^3 since there are three Pivots.

$$\begin{bmatrix} 1 & 0 & 0 \\ 0 & 1 & 0 \\ 0 & 0 & 1 \end{bmatrix}\begin{bmatrix} 0 \\ 0 \\ 0 \end{bmatrix} = 0$$

R^3 stands for Three Dimensions, thus three Pivots. Each pivot represents a Dimension.

This Linear Independency states that a Mapping of Matrix T of $R^n \rightarrow R^m$ is one to one since there is only one solution for a given b. For example: Tx = b

$$\begin{bmatrix} 1 & 0 & 0 \\ 0 & 1 & 0 \\ 0 & 0 & 1 \end{bmatrix}\begin{bmatrix} x \\ y \\ z \end{bmatrix} = \begin{bmatrix} 1 \\ 2 \\ 3 \end{bmatrix}$$ has one solution: $\begin{bmatrix} x \\ y \\ z \end{bmatrix} = \begin{bmatrix} 1 \\ 2 \\ 3 \end{bmatrix}$

Due to there being only one solution for each variable the Matrix is one to one.

Linear Dependent Matrix:

Suppose that there are three Vectors and two of them are a related.

$$\begin{bmatrix} 1 & 2 & 3 \\ 2 & 4 & 6 \\ 1 & 1 & 3 \end{bmatrix} = \textbf{Matrix}$$

Notice that the second Row is twice the first Row. This means that the Vectors at Row 1 and Row 2 are related which is an indication that the Matrix is Linearly Dependent.

Using ratios, multiplication, division, and addition, the Matrix above is simplified in the form:

$$\begin{bmatrix} 1 & 0 & 3 \\ 0 & 1 & 0 \\ 0 & 0 & 0 \end{bmatrix} = \textbf{Matrix}$$

Which means that if the matrix is set equal to zero there is the following:

The Matrix on the right Span R^2 since there are only two Pivots.

$$\begin{bmatrix} 1 & 0 & 3 \\ 0 & 1 & 0 \\ 0 & 0 & 0 \end{bmatrix} \begin{bmatrix} 0 \\ 0 \\ 0 \end{bmatrix} = 0$$

Mapping of Matrix T of $R^n \rightarrow R^m$ is onto since there is at least one solution for a given b. For example: Tx = b

$1x + 3z = 0$ and $1y = 0$

then

$X = -3z$ and $y = 0$ then

With z being the Free Variable

z can be any number

$$\begin{bmatrix} x \\ y \\ z \end{bmatrix} = z \begin{bmatrix} -3 \\ 0 \\ 1 \end{bmatrix}$$

Solutions Graphed

An example of a Unique Solution is:

$$\begin{bmatrix} 1 & 0 & 0 \\ 0 & 1 & 0 \\ 0 & 0 & 1 \end{bmatrix} \begin{bmatrix} x \\ y \\ z \end{bmatrix} = \begin{bmatrix} 1 \\ 2 \\ 3 \end{bmatrix} \text{ has one solution: } \begin{bmatrix} x \\ y \\ z \end{bmatrix} = \begin{bmatrix} 1 \\ 2 \\ 3 \end{bmatrix}$$

An example of Infinite Number of Solutions is:

$$\begin{bmatrix} x \\ y \\ z \end{bmatrix} = z \begin{bmatrix} -3 \\ 0 \\ 1 \end{bmatrix}$$ any time there are Free Variables there

is an Infinite Number of Solutions. Z can be any number.

An example of No Solution is:

$$\begin{bmatrix} 1 & 0 & 3 \\ 0 & 1 & 0 \\ 0 & 0 & 0 \end{bmatrix} \begin{bmatrix} 0 \\ 0 \\ 1 \end{bmatrix}$$ It impossible for 0z to equal 1 since there is

a multiplication by zero. This is an example of a Matrix
that is not consistent and thus have no solution.

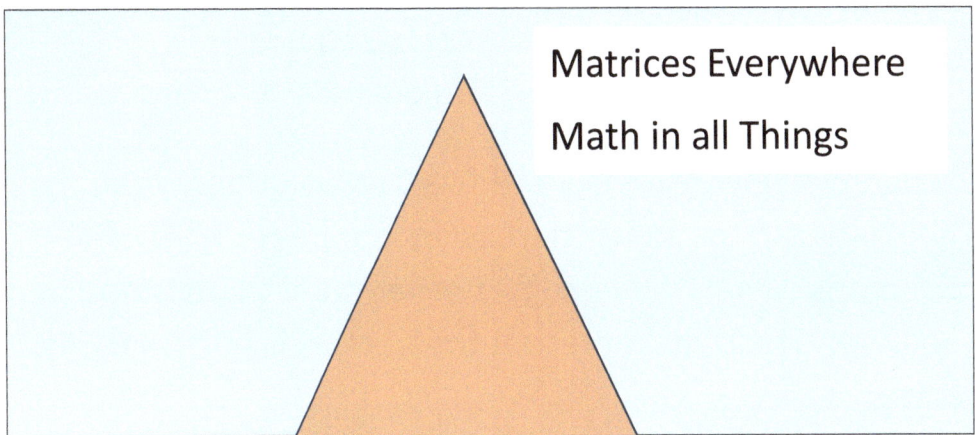

Matrices Everywhere

Math in all Things

Example of solutions:

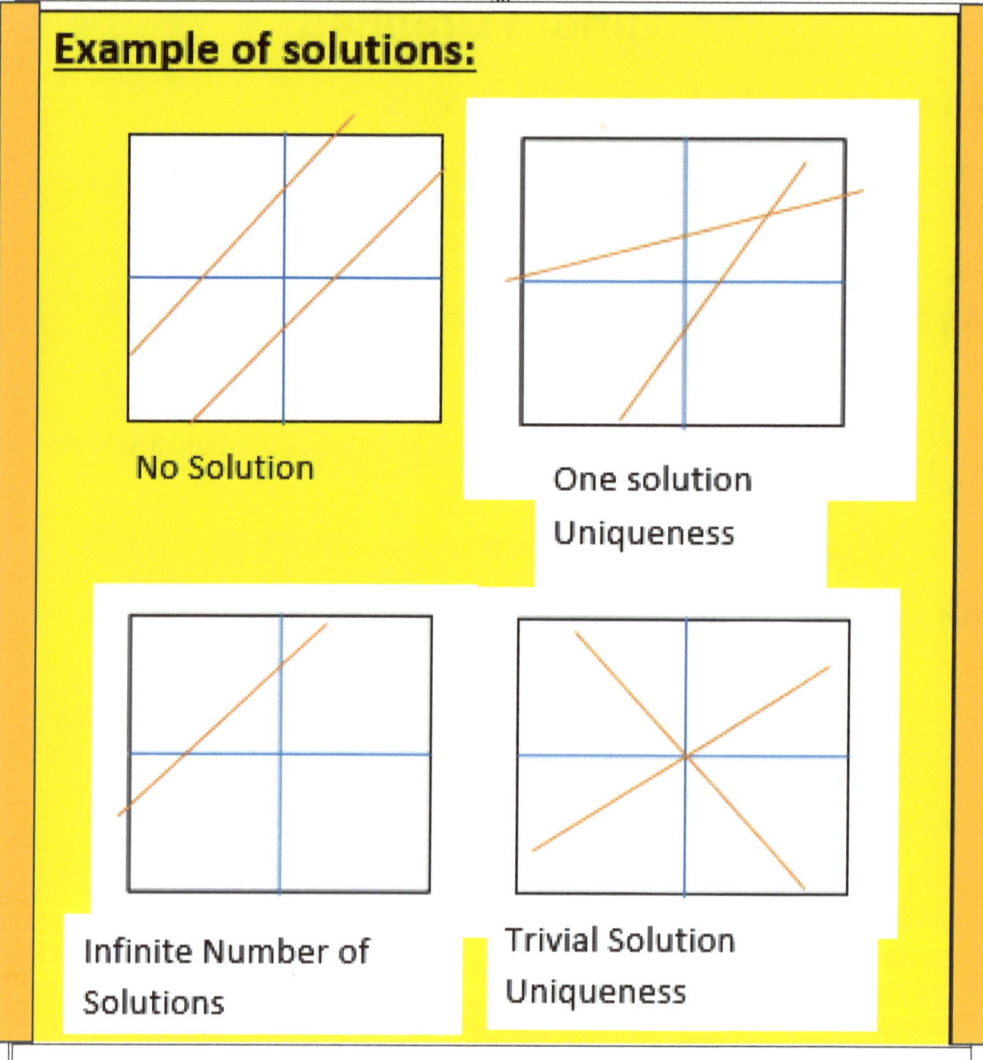

No Solution

One solution
Uniqueness

Infinite Number of
Solutions

Trivial Solution
Uniqueness

When two lines Intersect there is One Solution. When the two lines are the same there is an Infinite Number of Solutions. When the two lines never meet there is no Solution. The Lines represent the Vectors inside Matrices.

Subspace

A Subspace is formed as a result of Basis Vectors that when added together generates the many possible States and Wave Functions. A Subspace V W contains the zero Vector and linear combinations of V W such as:

$V + W$, $V - W$, $W - V$, cW, cV , and 0

A Subspace is a closed region where only addition, subtraction, the zero state, and multiples of its Basis Vectors are allowed. The Basis Vectors in the above example is V and W while c is a Scalar multiple.

The Entire Universe can be viewed as a Subspace where the collection of all the individual Wave Functions comprising the Basis, among the Matter and Energy in the Cosmos, leads to all Natural Phenomena and to all objects that we can see, feel, or detect. The Cosmos is literally a Sea of a Quantum Field where the Basis Vectors generate our Space Time Reality and all the Dimensions that holds things in place. That is why Quantum Mechanics is the study of everything, and Matrices and Linear Algebra are key subjects in Science, in an attempt to discover the codes that forms the Universe whether it is a simulation or not. All is measured, calculated, and can be understood.

A Subspace of Vector Space V is a subset of W of V if the following:

-The zero vector is in W:

$$0 = 0W_1 + 0W_2$$

-W is closed under addition:

If A and B is in W:

$$A = cW_1 + cW_2$$

$$B = dW_1 + dW_2$$

$$A + B = (c + d)W_1 + (c + d)W_2 = E$$

With E also being in W

-W is closed under multiplication of scalars C:

$$W = W_1 + W_2$$

$$cW = cW_1 + cW_2 = D$$

With D also being in W

All Basis Vectors are Linearly Independent and from these Vectors a Subspace is formed.

Eigenvector and Eigenvalue

Suppose that we have the Matrix:

$$\begin{bmatrix} 4 & 3 \\ 2 & 9 \end{bmatrix} = \textbf{Matrix}$$

The Determinant of the Matrix represents the Area that Vector 4x 3y and 2x 9y makes being these two vectors two sides of a Parallelogram.

The Determinant equation is for a general Matrix:

$$\begin{bmatrix} a & b \\ c & d \end{bmatrix} = \textbf{Matrix} \quad \textbf{Determinant = ad – bc}$$

Now suppose that there is an Operator that when Operating on a Function is equal to that Function times a Number.

(Operator)(function) = (Number)(Function)

The Number represents an Eigenvalue and the Function is an Eigenvector. There is an Eigenvector for each Eigenvalue. In Quantum Mechanics an Eigenvalue represents a Quantum State, and an Eigenvector represents a Wave Function in Space. A collection of Wave Functions leads to Space, which means that Eigenvectors forms a Basis.

It is possible to find Eigenvalues for a Matrix by doing the following. The desired result is Linearly Independent Eigenvectors, and Eigenvalues that represent an entire set Quantum States enclosed on itself leading to an Area of zero, which means that the Determinant must also equal zero.

The equation for that statement is:

$$A(\textbf{v}) = \lambda(\textbf{v})$$

Where A is the Operator, v the Eigenvector, and λ the Eigenvalue representing a Quantum State.

Which then leads to:

$$A(\textbf{v}) - \lambda(\textbf{v}) = 0$$

And we can solve for λ by using the Determinant which must also equal to zero for the Equation above.

$$Det((A-\lambda)v) = 0$$

Which leads to the following Matrix:

$$\begin{bmatrix} a - \lambda & b \\ c & d - \lambda \end{bmatrix} = \textbf{whose Determinant is zero.}$$

Using the Matrix shown earlier:

$$\begin{bmatrix} 4 & 3 \\ 2 & 9 \end{bmatrix} = \text{Matrix}$$

Then

$$\begin{bmatrix} 4 - \lambda & 3 \\ 2 & 9 - \lambda \end{bmatrix} = \text{Matrix}$$

Whose determinant must equal zero to find Eigenvalues.

Determinant = ad − bc

Then,

$(4-\lambda)(9 - \lambda) - 6 = 0$

Foiling:

$36 - 4\lambda - 9\lambda + \lambda^2 - 6 = 0$

$30 - 13\lambda + \lambda^2 = 0$

Which simplifies to:

$(\lambda - 3)(\lambda - 10) = 0$

With Eigenvalues equal to: $\lambda = 3$ and 10

So that means that there are two Quantum States in this Sea of Wave Functions.

To find the Eigenvectors the equation used is:

$$A(v) - \lambda(v) = 0$$

So for Quantum State $\lambda = 3$

$$\begin{bmatrix} 4 - \lambda & 3 \\ 2 & 9 - \lambda \end{bmatrix} = \begin{bmatrix} 0 \\ 0 \end{bmatrix}$$

Then,

$$\begin{bmatrix} 4 - 3 & 3 \\ 2 & 9 - 3 \end{bmatrix} = \begin{bmatrix} 0 \\ 0 \end{bmatrix}$$

Then,

$$\begin{bmatrix} 1 & 3 \\ 2 & 6 \end{bmatrix} = \begin{bmatrix} 0 \\ 0 \end{bmatrix}$$

Which simplifies to:

$$\begin{bmatrix} 1 & 3 \\ 0 & 0 \end{bmatrix} = \begin{bmatrix} 0 \\ 0 \end{bmatrix}$$

Which means that:

$1x + 3y = 0$

$x = -3y$

the Eigenvector for $\lambda = 3$ is:

$$v = y \begin{bmatrix} -3 \\ 1 \end{bmatrix} \quad \text{leading to} \quad \begin{bmatrix} -3 \\ 1 \end{bmatrix}$$

To find the Eigenvectors the equation used is:

$$A(\mathbf{v}) - \lambda(\mathbf{v}) = 0$$

So for Quantum State $\lambda = 10$

$$\begin{bmatrix} 4 - \lambda & 3 \\ 2 & 9 - \lambda \end{bmatrix} = \begin{bmatrix} 0 \\ 0 \end{bmatrix}$$

Then,

$$\begin{bmatrix} 4 - 10 & 3 \\ 2 & 9 - 10 \end{bmatrix} = \begin{bmatrix} 0 \\ 0 \end{bmatrix}$$

Then,

$$\begin{bmatrix} -6 & 3 \\ 2 & -1 \end{bmatrix} = \begin{bmatrix} 0 \\ 0 \end{bmatrix}$$

Which simplifies to:

$$\begin{bmatrix} 1 & -1/2 \\ 0 & 0 \end{bmatrix} = \begin{bmatrix} 0 \\ 0 \end{bmatrix}$$

Which means that:

1x - (1/2)y = 0

x = (1/2)y

the Eigenvector for $\lambda = 10$ is:

$\mathbf{v} = y \begin{bmatrix} 1 \\ 2 \end{bmatrix}$ leading to $\begin{bmatrix} 1 \\ 2 \end{bmatrix}$ since y is twice x

Now Matrix= $\begin{bmatrix} 4 & 3 \\ 2 & 9 \end{bmatrix}$ has two Eigenvalues whose Eigenvectors are Linearly Independent thus forming a Basis from where other Wavefunctions can be formed. By forming a new matrix with both Eigenvectors we have:

$$\begin{bmatrix} -3 & 1 \\ 1 & 2 \end{bmatrix} = \text{Eigenvectors lined up}$$

Which simplifies into:

$\begin{bmatrix} 1 & 0 \\ 0 & 1 \end{bmatrix} = $ which proofs that the Eigenvectors are Linearly Independent since only the Trivial Solution exists for the equation Tx = 0

$$\begin{bmatrix} 1 & 0 \\ 0 & 1 \end{bmatrix} = 0$$

Where x = $\begin{bmatrix} 0 \\ 0 \end{bmatrix}$ thus forming a Basis from where other

A Maze of Wavefunctions!

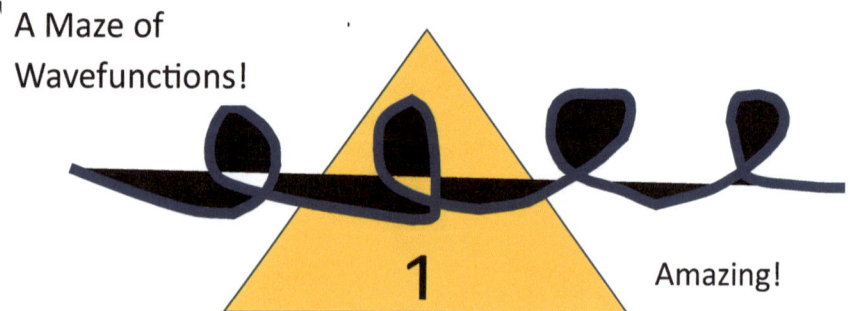

1

Amazing!

Proportions and Ratios

Everything in the Cosmos exists in Proportions to other things. There is the right amount of Oxygen in the air for us to breathe and anything more or less would be a problem. There is the right amount of water to drink every day, there is the right amount of exercise, the right Proportion of red meat to other forms of food to intake weekly. The Ratios and Proportions are what drives the Symphony of Universal Harmony. The Universe is like a Music with flows and with bursts of Energy whether in the Stars or Black Holes, or between Neurons in our Brains. There is a Flux of events with Energy, Density, Pressure, and these disturbances have Ratios and Proportions, and if this Symphony had been at a different Frequency it would amount to an entirely different reality than what we can perceive. We perceive by absorbing incoming Light, Temperature, Air, and Sound. The Light allows us to see with our eyes, the Temperature can be felt with our skin, the Air we absorb with the Lungs to breathe, and Sound goes into our ears allowing hearing. All of these exist in Ratios and Proportions with given Frequencies and Energies. Everything is either a Wave or a Circle with Cycles and Frequencies.

If the Ratio of Matter A and Matter B is that for every Matter A there are 3 Matter B, we have a Proportion:

Matter A $\propto$ 3 Matter B

So, by following the same Proportion, if now we have 4 Matter A, how many Matter B do we have?

4 Matter A $\propto$? Matter B

We can solve it by doing Ratios since the Proportions are kept the same.

Matter A $\propto$ 3 Matter B

4 Matter A $\propto$? Matter B

Which then becomes:

A $\propto$ 3 B

4 A $\propto$? B

Solving for ? we get:

$$\frac{A}{4A} = \frac{3B}{?B}$$

?B = 4(3)B

Which means that there are 12 Matter B when there are 4 Matter A.

Ratios are in everything and inside Molecules which comprises the Matter. The right amount of Oxygen in Proportion to Hydrogen forms Water. There are 2 Atoms of Hydrogen and 1 Atom of Oxygen in a Water Molecule. When it comes to Mass, the Oxygen Atom is heavier than Hydrogen and most of Water in Weight is Oxygen despite there only being a single Atom of Oxygen for every one of its Molecules. The Universe is a series of Mathematical Combinations, whether in Wave Form such as in Light or in Particles with Amplitude and Wavelength, or in building blocks at the Quantum Level where Atoms arrange themselves to Form Molecules such as the DNA. These combinations are the Signature of Life. What makes a dog different from a man is the DNA with a set of codes that when read line by line generates a unique phenomenon, and each person has a DNA that makes an individual unique although very related to other living beings. What exactly reads these lines of code in DNA and in general Matter is a mystery. Combinations, Ratios, and Proportions sum up to generate our Reality and what and how we perceive this Reality.

Kepler's Three Laws

On the shoulder of giants is where Scientists sit by learning from what Scientists in the past discovered which is recorded in books, and then refining this knowledge with better explanations and by adding more discoveries to them with regular updates. Kepler's Three Laws, for example, are in perfect agreement with Newton's conception of Gravity. Kepler's Three Laws state the following:

1....The orbit of Planets are elliptical.

2...The Planets sweep equal areas in their orbit in equal amount of time.

3...The square of a Planet's Orbital Period is Proportional to the cube of the Planet's Average Distance to the Sun.

In the next page there is a proof for Kepler's Third Law. These discoveries were made from observations recorded by Tycho Brahe of the motion of Planets, the Moon, and the Sun performed before the invention of the Telescope. A pattern of these motions was noticed, but only later would Equations be used into describing the motion with more accuracy and precision.

Proof:

Force of Gravity between the Sun and a Planet is:

$$Force = \frac{GMm}{r^2}$$

Where G is the Gravitational Constant, M is the Mass of the Sun, m the Mass of a Planet, and r the Distance between the Center of the Sun and the Center of the Planet.

By setting the Force of Gravity equal to the Centripetal Force that keeps the Planet moving in an Orbit around the Sun:

$$\frac{GMm}{r^2} = \frac{mv^2}{r}$$

Simplifying:

$$\frac{GM}{r} = \frac{v^2}{1}$$

V is the Speed of the Orbit

Assuming just for convenience that the Orbit of Planets is a Circle whose Circumference is $2\pi r$ with r being the Average Distance between the Planet and the Sun:

V which is the Speed that the Planet Orbits the Sun is equal to:

$$V = \frac{2\pi r}{T}$$ Where T is the Orbital Period

By plugging this result into the previous Equation:

$$\frac{GM}{r} = \left(\frac{2\pi r}{T}\right)^2$$

Which simplifies to:

$$\frac{GM}{4\pi^2} = \frac{r^3}{T^2}$$

Which means the following:

$$T^2 \frac{GM}{4\pi^2} = \frac{r^3}{1}$$

Then the Proportionality Relation is:

$$T^2 \propto r^3$$

The square of a Planet's Orbital Period is Proportional to the cube of the Planet's Average Distance to the Sun.

Proportions and the Third Law:

Let us say that Planet Earth is now in a different Solar System but with the same length of Year and Distance from the Star. It is possible to know the Orbital Period and Distance between any Planet and the Sun if we know at least one of the two information.

Using:

$$T^2 \propto r^3 \text{ for Earth is } (1 year)^2 \propto (1 AU \ Distance)^3$$

Where 1 AU stands for 1 Astronomical Unit which is the Distance between Earth and the Sun.

Let us say that Planet Zir is 1.5AU away from the Sun. Using Proportionality its Orbital Period is:

$$T^2 \propto 1.5^3$$

$$T = \sqrt{1.5^3} = 1.84 \ Years$$

Now if we knew that Planet Zir has an Orbital Period of 1.84 years, it would be possible to find its Distance with:

$$(1.84)^2 \propto R^3$$

$$R = \sqrt[3]{1.84^2} = 1.50 \text{ AU}$$

This only a small example on how Proportionality can be useful in Physics.

Intensity and Amplitude:

Another example of Proportionality is Intensity and Amplitude of an emitted Wave. The Source of a Wave emits power per square meter that after travelling a Distance gets less intense by obeying the inverse square law. The Frequency and Wavelength of the Wave remains the same, but the Amplitude gets lower. The Proportions are shown below:

$$\text{Intensity} \propto (Amplitude)^2$$

$$\text{Intensity} \propto \left(\frac{1}{Distance}\right)^2$$

The Frequency of the Wave remains the same at any Distance, and if that was not the case, Radio Stations would not be able to transmit their channels. Thankfully the Intensity and Amplitude drops with greater Distance. This means that the same Radio Frequency in Dallas belongs to a different Radio Station than the one with the same Frequency at Houston. This fact eliminates interference and difficulties broadcasting Images and Sound.

Proportions, Intensity, and Amplitude:

Suppose that there is a Source of Wave that at a Distance of 4.0 m has an Intensity of 50 w/m^2 and Amplitude 1.0 m.

That means that using:

Intensity $\propto (\dfrac{1}{Distance})^2$

At a Distance of 8.0 m it will emit:

$$50 \propto (\tfrac{1}{4})^2$$

$$I \propto (\tfrac{1}{8})^2$$

$$\dfrac{50}{I} : \dfrac{(\tfrac{1}{4})^2}{(\tfrac{1}{8})^2} \qquad \dfrac{50}{I} : \dfrac{8^2}{4^2} \qquad \dfrac{I}{50} : \dfrac{4^2}{8^2}$$

Solving for I we get: **I = 12.5 W/m^2**

And the Amplitude at 8.0 m will be using:

Intensity $\propto (Amplitude)^2$

$$\dfrac{50}{12.5} : \dfrac{1^2}{A^2} \qquad A = \sqrt{\dfrac{12.5}{50}} = \textbf{0.5 m}$$

Proofs for Inverse Square Law:

Whenever there is a Source of a Wave, the emitted Wave or Pulse spreads around and from the Source with circles in two Dimensions with a Radius that is equal to the Distance from the Source. The Waves then expand in concentric circles that in three Dimensions are shaped like Spheres. The Physicist called Isaac Newton reasoned that since the Intensity of a Wave is higher the closer an observer is to the Source, the Intensity must be inversely proportional to the Surface Area of a Sphere with Radius equal to the Distance from the Source. Since the Surface Area of a Sphere is $4\pi r^2$, Intensity is inversely proportional to r^2:

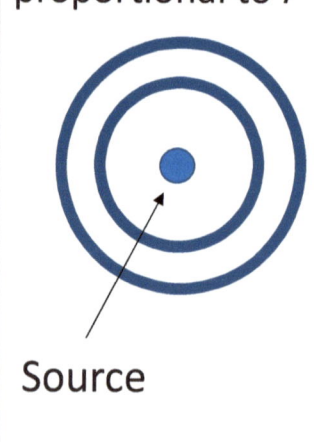

Source

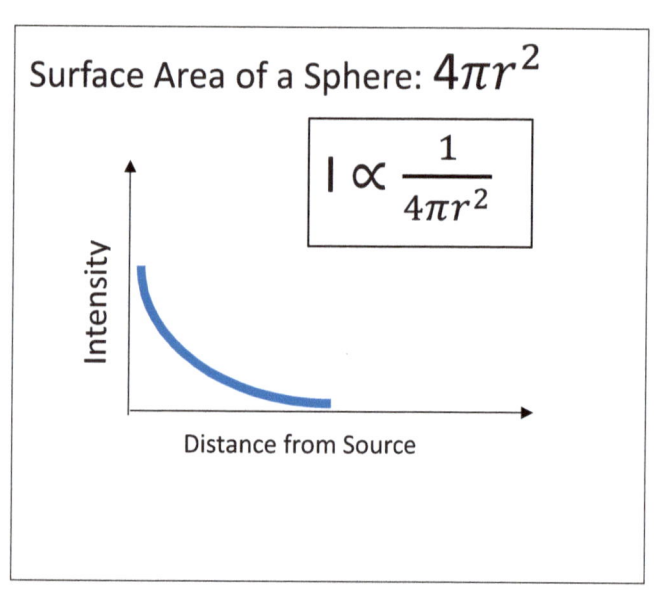

Surface Area of a Sphere: $4\pi r^2$

$$I \propto \frac{1}{4\pi r^2}$$

Intensity

Distance from Source

<u>Adding Vectors:</u>

Vectors are numbers with Directions while Scalars are just numbers. Examples of Vectors are Forces, Displacement, Weight, Velocity and examples of Scalars are Temperature, Speed, Distance, and Mass. Velocity is Speed with Direction and Displacement is Distance with Direction. When adding Vectors, Mathematicians arrive at the Resultant which is the Sum of Vectors. In order to add Vectors, the Components of each Vector are added leading to new Components which comprise the Resultant.

Example 1:

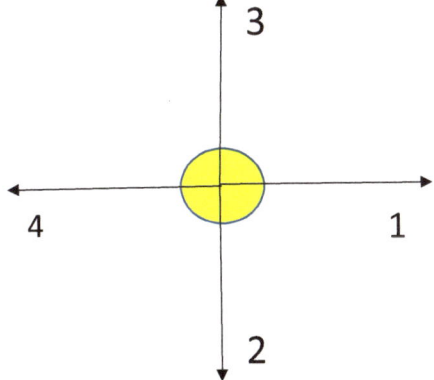

Here are four Vectors:

3 Up, 2 Down, 1 Right, and 4 Left

The Components x and y are added individually:

X component	Y component
-4	3
1	-2
Total	
-3	1

The Resultant Vector has components:

X = -3   and Y = 1

Drawing these Components we get:

3 to the Left and 1 Up: head to tail

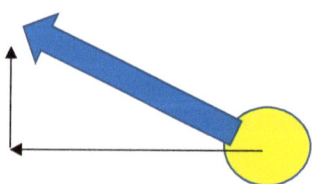

The Hypotenuse is the Resultant.

Hypotenuse = $\sqrt{3^2 + 1^2} \approx 3.16$

Direction of Vectors:

All Vectors point in a Direction with respect to the Positive x-axis. The x axis is conceptualized horizontally, and the y axis is vertically. The Positive x axis is towards the right according to the convention.

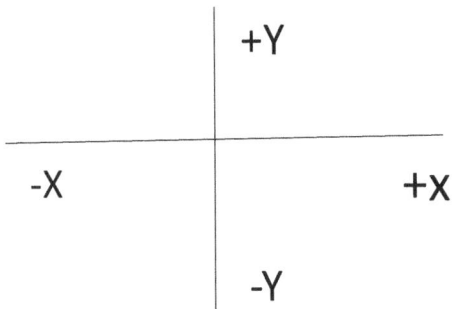

Examples of Directions:

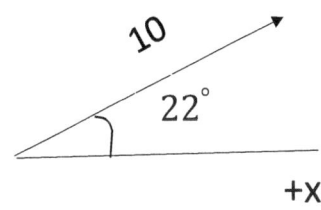

10 at $22°$ from the Positive x axis

Example 2:

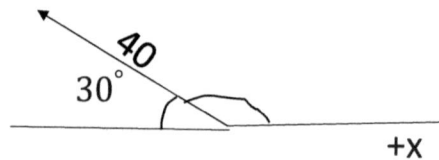

40 at $150°$ from the Positive x axis.

In this case 180 -30 gives the direction from the Positive x axis.

Example 3:

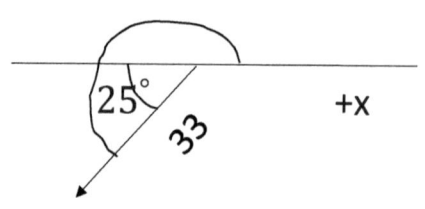

33 at $205°$ from the Positive x axis.

In this case 180 + 25 gives the direction from the Positive x axis.

Example 4:

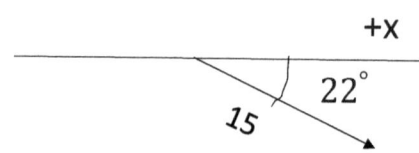

15 at $-22°$ from the Positive x axis.

In this case the negative of 22 gives direction from Positive x axis.

Quantum Mechanics Matrices:

Suppose that there is a Particle in Space and the graph below shows the Probabilities of the Particle being in certain areas of Space.

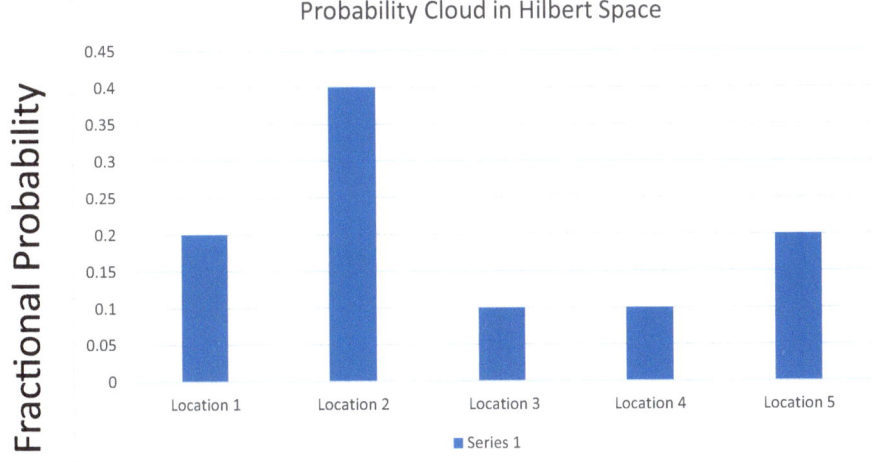

There is a 20% chance of the Particle being in Location 1, 40% chance of being in Location 2, 10% chance of being in Location 3, 10% chance of being in Location 4, and 20% chance of being in Location 5. There is 100% chance of the Particle being in Space. Particles are Waves of Probability, which we call a Probability Cloud. The Particle is not the cloud, and what it is a mystery. Theories predict that the Probabilities are the several locations of the Particle in Parallel Universes. The Particle exists in a Wave of Probabilities like a Gaussian Curve. Only when observed does the Particle appear to be in one of the Locations. The act of Observing interferes with the Particle.

Table of probabilities:

Space	Probability
Location 1	0.2
Location 2	0.4
Location 3	0.1
Location 4	0.1
Location 5	0.2

Into Matrix Form:

$$\begin{pmatrix} 0.2 \\ 0.4 \\ 0.1 \\ 0.1 \\ 0.2 \end{pmatrix} \quad \text{Into} \quad \longrightarrow \quad \begin{pmatrix} \sqrt{0.2} \\ \sqrt{0.4} \\ \sqrt{0.1} \\ \sqrt{0.1} \\ \sqrt{0.2} \end{pmatrix}$$

Now there is:

$$|\varphi> = \begin{pmatrix} \sqrt{0.2} \\ \sqrt{0.4} \\ \sqrt{0.1} \\ \sqrt{0.1} \\ \sqrt{0.2} \end{pmatrix} \quad <\varphi| = \begin{pmatrix} \sqrt{0.2} & \sqrt{0.4} & \sqrt{0.1} & \sqrt{0.1} & \sqrt{0.2} \end{pmatrix}$$

The sum of the Probabilities is equal to 1 according to the equation for the Gaussian Curve:

$$\int_{Location\ 1}^{location\ 5} <\varphi|\varphi> dx\ = 1$$

The Matrix Multiplication:

$$<\varphi|\varphi> \ = 1 = |\ \varphi><\varphi|$$

$$\begin{pmatrix} \sqrt{0.2} \\ \sqrt{0.4} \\ \sqrt{0.1} \\ \sqrt{0.1} \\ \sqrt{0.2} \end{pmatrix} \begin{pmatrix} \sqrt{0.2} & \sqrt{0.4} & \sqrt{0.1} & \sqrt{0.1} & \sqrt{0.2} \end{pmatrix} = 1$$

It is not possible to state with 100% certainty where the Particle is, but there is a 100% certainty that it is somewhere within the Hilbert Space of our Universe. In order for the Matrix Multiplication to equal 1 the square root was added to the Probabilities.

Expectation Value:

The Expectation Value is the most likely state of a Particle. The Expectation value of Position is the most likely Position. It is equal to:

Expectation Value = $< \varphi|x|\varphi >$

Matrix x is equal to the Positions:

$$X = \begin{pmatrix} 1 & 0 & 0 & 0 & 0 \\ 0 & 2 & 0 & 0 & 0 \\ 0 & 0 & 3 & 0 & 0 \\ 0 & 0 & 0 & 4 & 0 \\ 0 & 0 & 0 & 0 & 5 \end{pmatrix}$$

Quantum Mechanics is the study of Matrices, Space, Probabilities, and the uncertain aspects of Reality at the Subatomic level.

$$< \varphi|x|\varphi > = \begin{pmatrix} \sqrt{0.2} & \sqrt{0.4} & \sqrt{0.1} & \sqrt{0.1} & \sqrt{0.2} \end{pmatrix} \begin{pmatrix} 1 & 0 & 0 & 0 & 0 \\ 0 & 2 & 0 & 0 & 0 \\ 0 & 0 & 3 & 0 & 0 \\ 0 & 0 & 0 & 4 & 0 \\ 0 & 0 & 0 & 0 & 5 \end{pmatrix} \begin{pmatrix} \sqrt{0.2} \\ \sqrt{0.4} \\ \sqrt{0.1} \\ \sqrt{0.1} \\ \sqrt{0.2} \end{pmatrix}$$

Which means the Following:

1(0.2) + 2(0.4) + 3(0.1) + 4(0.1) + 5(0.2) = 2.70

The Expectation Value is 2.70 between Positions 2 and 3.

Linear Operators and Hilbert Space:

For the example shown before, the Eigenvalues are the following:

$\sqrt{0.2} \; for \; x = 1$ There is an Eigenvalue for each of the Eigenvectors that together form the

$\sqrt{0.4} \; for \; x = 2$ Hilbert Space which is a collection of

$\sqrt{0.1} \; for \; x = 3$ Wave Functions.

$\sqrt{0.1} \; for \; x = 4$

$\sqrt{0.2} \; for \; x = 5$

Eigenvectors and Eigenvalues:

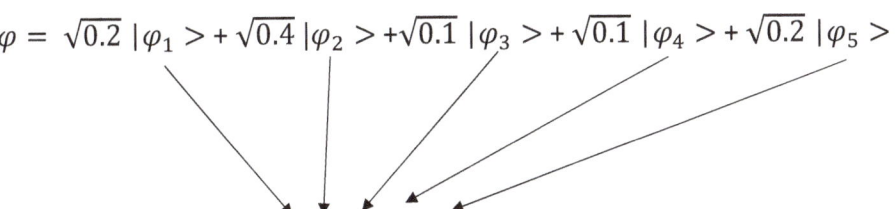

$$\varphi = \sqrt{0.2} \; |\varphi_1 > + \sqrt{0.4} \; |\varphi_2 > + \sqrt{0.1} \; |\varphi_3 > + \sqrt{0.1} \; |\varphi_4 > + \sqrt{0.2} \; |\varphi_5 >$$

Eigenvectors with their Eigenvalues

The Absolute Value of an Eigenvalue Squared is equal to the Probability of the Particle being in that Quantum State. Example shown below:

$|(\sqrt{0.2})^2| = 0.2 \; the \; Probability \; of \; Particle \; being \; in \; position \; x = 1$

Finding the Normalization Constant:

Suppose that there is a Particle in Space and the graph below shows the Probabilities of the Particle being in certain areas of Space.

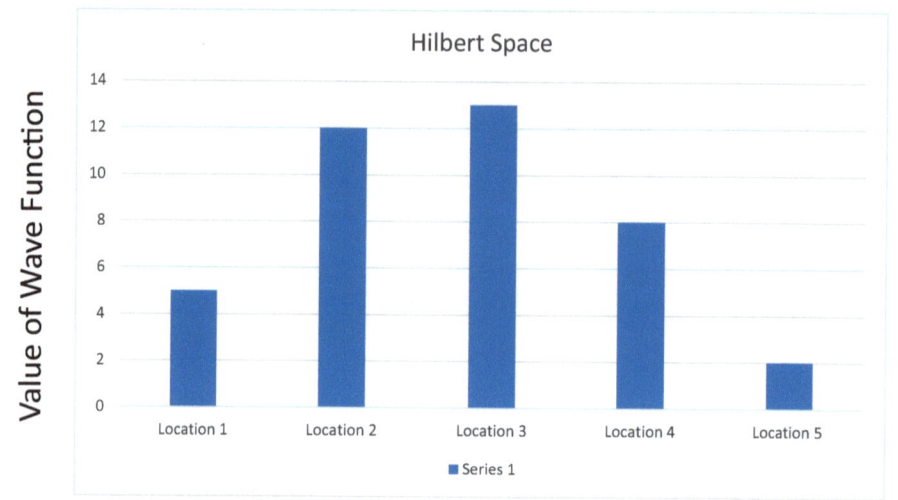

The Table for the Graph above is:

Location	Wave Function Value
1	5
2	12
3	13
4	8
5	2

Notice however that the Wave Function value does not add up to 100%, which means that there is a need to find a Normalization Constant.

It is necessary to find A such that the bottom equation will work:

$$A^2 \int_{Location\ 1}^{location\ 5} <\varphi|\varphi> \, dx = 1$$

And 5+12+13+8+2 = 40

So, by diving the Wave Functions by 40 and taking the Square Root:

Location	Eigenvalue
1	$\sqrt{\dfrac{5}{40}}$
2	$\sqrt{\dfrac{12}{40}}$
3	$\sqrt{\dfrac{13}{40}}$
4	$\sqrt{\dfrac{8}{40}}$
5	$\sqrt{\dfrac{2}{40}}$

The Eigenvectors and Eigenvalues are:

$$\varphi = \sqrt{\frac{5}{40}} \,|\varphi_1> + \sqrt{\frac{12}{40}} \,|\varphi_2> + \sqrt{\frac{13}{40}} \,|\varphi_3> + \sqrt{\frac{8}{40}} \,|\varphi_4> + \sqrt{\frac{2}{40}} \,|\varphi_5>$$

Where the Probability Density Equation is:

$$A^2 < \varphi|\varphi > = \frac{5}{40} < \varphi_1 \,|\varphi_1> + \frac{12}{40} < \varphi_2 |\varphi_2> + \frac{13}{40} < \varphi_3|\varphi_3> +$$

$$\frac{8}{40} < \varphi_4|\varphi_4> + \frac{2}{40} < \varphi_5|\varphi_5>$$

The sum of the Eigenvalues Squared leads to a Probability of 100% that the Particle is somewhere in the Hilbert Space.

$$\frac{5}{40} + \frac{12}{40} + \frac{13}{40} + \frac{8}{40} + \frac{2}{40} = 1 \text{ or } 100\%$$

Which means that A the Normalization Constant is:

$$A = \sqrt{\frac{1}{40}}$$ Allowing the bottom equation to be true:

$$A^2 \int_{Location\ 1}^{location\ 5} < \varphi|\varphi > dx = 1$$

And by placing the value of A into the equation:

$$(\sqrt{\frac{1}{40}})^2 \int_{Location\ 1}^{location\ 5} < \varphi|\varphi > dx = 1$$

The Superposition in this case has 5 possible states for the Particle. Until the Particle is measured it can exist in any of the 5 states. That means that there are 5 Parallel Universes where the Particles exist before being measured thus accounting for the 5 cases.

Proof of Taylor's Series:

Taylor Series is an Infinite Sum that that defines a Function. The addition of an Infinite Number of Terms which converges to a Single Function Value at a point in the Graph. It can be proved through the following:

Power series is:

$F(x) = \sum_{n=0}^{\infty} a_0 x^n = a_0 + a_1 x + a_2 x^2 + a_3 x^3 + \ldots$

When x = 0, $F(x) = a_0$

$F'(x) = a_1 + 2a_2 x + 3a_3 x^2 + \ldots$

When x = 0, $F'(x) = a_1$

$F''(x) = 2a_2 + 6a_3 x + \cdots$

When x = 0, $F''(x) = 2a_2$

Which means that: $\dfrac{F''(x)}{2} = a_2$

Which leads to the final conclusion:

$F(x) = \sum_{n=0}^{\infty} \dfrac{f^n(a)(x)^n}{n!}$

Taylor Series are used to definite a great Number of Functions as an Infinite Sum. These Series can also prove Euler's Equation that $e^{ix} = cosx + isinx$

Power Series are centered at x = 0 while Taylor Series can be centered anywhere in the Graph.

Infinite Sums and the Hilbert Space:

Quantum Physics states that Particles and Matter are comprised of a collection of Wave Functions that when added together generates the Reality that we see. Power and Taylor Series follows the Mathematical Concept that all Functions in Space can be derived from an Infinite Sum that converges to a certain value that represents the Function at that value. This revolves around the idea that by finding a Pattern from the Derivatives in the Sum, the Coefficients are then found in the way that it becomes possible to make the Sum better fit as a description of Reality. Since the Sum goes into Infinity, each term adds more accuracy to the result and in the Limit of terms into Infinity it is possible to arrive at the Function which is now viewed as a Sum. The Mathematics is a beautiful representation that from Infinity it is possible to arrive at Functions, which could be Wave Functions that added on top of each other generates all that we see as a product of the Hilbert Quantum Field Space.

Proof of the Equation that $e^{ix} = cosx + isinx$:

Using Taylor Series:

$$F(x) = \sum_{n=0}^{\infty} \frac{f^n(a)(x)^n}{n!}$$

The Equation e^{ix} becomes the following:

If we center our approximation at zero and take the derivatives:

$F(x) = e^{ix}$

$f'(x) = ie^{ix}$

$f''(x) = -1e^{ix}$
$f'''(x) = -ie^{ix}$

$f''''(x) = e^{ix}$

Plugging zero for x:

F(0) = 1

$f'(0) = i$

$f''(0) = -1$

$f'''(0) = -i$

$f''''(0) = 1$

Which means that:

$$F(x) = \sum_{n=0}^{\infty} \frac{f^n(a)(x)^n}{n!}$$

Becomes

$$e^{ix} = \sum_{n=0}^{\infty} \frac{(i)^n x^n}{n!}$$

Using Taylor Series, Sin(x) is:

$$F(x) = \sum_{n=0}^{\infty} \frac{f^n(a)(x)^n}{n!}$$

The equation Sin(x) becomes the following:

If we center our approximation at zero and take the derivatives:

F(x) = Sin(x)

$f'(x) = Cos(x)$

$f''(x) = $ -Sin(x)

$f'''(x) = $ -Cos(x)

$f''''(x) = Sin(x)$

Plugging zero for x:

F(0) = 0

$f'(0) = 1$

$f''(0) = 0$

$f'''(0) = $ -1

$f''''(0) = 0$

Which means that:

$$F(x) = \sum_{n=0}^{\infty} \frac{f^n(a)(x)^n}{n!}$$

Becomes

$$Sin(x) = \sum_{n=0}^{\infty} \frac{(-1)^n x^{2n+1}}{(2n+1)!}$$

Only Odd Derivates provides a contribution to the Series.

Using Taylor Series, Cos(x) is:

$$F(x) = \sum_{n=0}^{\infty} \frac{f^n(a)(x)^n}{n!}$$

The equation Cos(x) becomes the following:

If we center our approximation at zero and take the derivatives:

$F(x) = Cos(x)$

$f'(x) = -Sin(x)$

$f''(x) = -Cos(x)$

$f'''(x) = Sin(x)$

$f''''(x) = Cos(x)$

Plugging zero for x:

$F(0) = 1$

$f'(0) = 0$

$f''(0) = -1$

$f'''(0) = 0$

$f''''(0) = 1$

Which means that:

$$F(x) = \sum_{n=0}^{\infty} \frac{f^n(a)(x)^n}{n!}$$

Becomes

$$Cos(x) = \sum_{n=0}^{\infty} \frac{(-1)^n x^{2n}}{(2n)!}$$

Only Even Derivates provides a contribution to the Series.

Thus, it can be concluded that:

$$\sum_{n=0}^{\infty} \frac{(i)^n x^n}{n!} = \sum_{n=0}^{\infty} \frac{(-1)^n x^{2n+1}}{(2n+1)!} + \sum_{n=0}^{\infty} \frac{(-1)^n x^{2n}}{(2n)!}$$

Which means that:

$$e^{ix} = cosx + isinx$$

This implies that both e and i in the Equation above can generate a Wave which is a combination of Sine and Cosine, being Sine the Imaginary Part and Cosine the Real Part of the Oscillation.

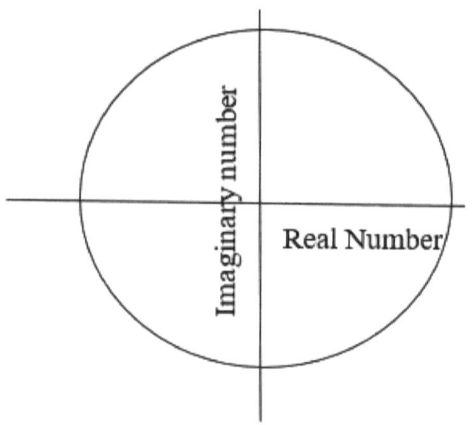

The imaginary i in front of the Sinx is due to the fact that the vertical axis in this imaginary number circle is imaginary while the horizontal is real.

Imaginary Numbers are as Real as it can get! Imaginary Numbers exist to make Mathematics work for Scientist to be able to understand the Logos Cosmos.

Addition of Taylor Series:

$$\text{Cos}(x) = 1 - \frac{x^2}{2!} + \frac{x^4}{4!} - \frac{x^6}{6!} + \frac{x^8}{8!} \ldots$$

$+$

$$i\text{Sin}(x) = i\left(x - \frac{x^3}{3!} + \frac{x^5}{5!} - \frac{x^7}{7!} + \frac{x^9}{9!} \ldots\right)$$

$=$

$$e^{ix} = 1 + ix - \frac{x^2}{2!} - i\frac{x^3}{3!} + \frac{x^4}{4!} + i\frac{x^5}{5!} \ldots$$

Thus, proving Euler's Equation:

$$e^{ix} = \cos x + i \sin x$$

And:

$$e^{i\pi} + 1 = 0 \quad \text{Euler's Identity}$$

Proof for Schrodinger's Equation:

The Schrodinger's Equation describes the Wave Function of a Particle with respect to Position and Time:

$$\frac{\partial}{\partial x}\psi(x,t) = \text{Schrodinger's Equation}$$

It has been proven earlier that $e^{ix} = cosx + isinx$ a representation of a wave.

Which means that $\psi(x,t) = Ae^{i(kx-wt)}$

That fact then leads to: $\frac{\partial}{\partial x}\psi(x,t) = \frac{\partial}{\partial x}(Ae^{i(kx-wt)})$

By taking the Derivative: $\frac{\partial}{\partial x}\psi(x,t) = ikAe^{i(kx-wt)}$

Since, $\psi(x,t) = Ae^{i(kx-wt)}$

Now,

$$\frac{\partial}{\partial x}\psi(x,t) = ik\,\psi$$

And,

$$k = \frac{2\pi}{\lambda} \qquad \text{wave number}$$

Then using de Broglie's formula:

$$k = \frac{2\pi P}{h}$$

Then using de Broglie's formula:

$$k = \frac{2\pi P}{h}$$

$$k = \frac{P}{\hbar}$$

$$\frac{\partial \psi}{\partial x} = i\frac{P}{\hbar}\psi$$

$$\frac{\hbar}{i}\frac{\partial \psi}{\partial x} = -P\psi$$

$-i\hbar\dfrac{\partial \psi}{\partial x} = P\psi$ the momentum section of the equation

The Time Portion of the Equation leads to:

$$\frac{\partial \psi}{\partial t} = \frac{\partial}{\partial t}(Ae^{i(kx-wt)})$$

$$= -iwAe^{i(kx-wt)}$$

$$= -iwA\psi$$

From: E = hf

$E = h\dfrac{w}{2\pi} = \hbar w$ since $w/(2\pi) = f$ w = Angular
Frequency

We now have:

$$\frac{\partial \psi}{\partial t} = -i\frac{E}{\hbar}\psi$$

Or,

$$-\frac{\bar{h}}{i}\frac{\partial\psi}{\partial t} = E\,\psi \qquad \text{the time portion of the equation}$$

The energy equation is:

$$E = \frac{P^2}{2m} + V(x, t)$$

$$E\psi(x,t) = (\frac{P^2}{2m} + V(x, t))\,\psi(x,t)$$

$$i\bar{h}\frac{\partial\psi(x,t)}{\partial t} = \frac{1}{2m}\left(\left(-i\bar{h}\frac{\partial\psi}{\partial x}\right)^2 + V(x, t)\right)\psi(x,t)$$

The final equation is:

$$i\bar{h}\frac{\partial\psi}{\partial t} = -\bar{h}^2\frac{\partial^2\psi}{2m\partial x^2} + V\psi$$

Energy Levels:

A good way to visualize Energy Levels is with Standing Waves. These type of Waves represent, for example, a vibrating string with its ends firmly tight. When the strings are allowed to vibrate, they move up and down rapidly and forming a Standing Wave.

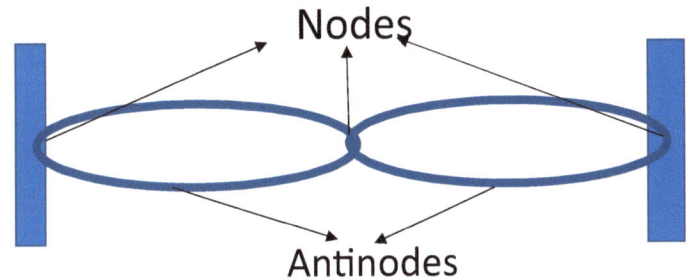

The picture above shows a Standing Wave. It has 3 Nodes and 3 Antinodes. One full Wavelength is comprised of 2 Antinodes as shown above. The Nodes are regions in the Standing Wave with no Amplitude, and the Antinodes are the bulges.

Strings vibrate according to their Energy. The greater the number of bulges or Antinodes, the greater is the Energy, and the smaller is its Wavelength. Smaller Wavelengths are the result of higher Frequency, and higher Energy:

Energy = hf

h is Planck's Constant and f is the Frequency

Since E = hf

And Using the Speed of Light as c, the equation $c = \lambda f$

where λ is Wavelength and f is Frequency, with $f = \dfrac{c}{\lambda}$

leads to the conclusion that $E = h\dfrac{c}{\lambda}$. The smaller the Wavelength the higher the Energy.

Standing Waves can serve as a map for Energy Levels:

No Energy, No Oscillation

First Energy Level, the Ground State.

Second Energy Level

Third Energy Level

Velocity of a Wave = (Wavelength)(Frequency)

If we assume that the Length between the two ends is L, then the Wavelengths for each Energy Level is:

$\lambda = 0 \quad f = 0$

$\lambda = 2L \quad f = v / (2L)$

$\lambda = L \quad f = v / (L)$

$\lambda = 2L/3 \quad f = 3v / (2L)$

Where v is the Speed of the Wave in the string going back and forth generating the Standing Wave.

The General Equation is:

$\lambda = \dfrac{2L}{n}$ and $f = v / \lambda$

Where n is the Energy Level. The Ground State is Energy Level 1

Particles have Energy due to the fact that they are Oscillations in Space. These Oscillations propagate in space leading to the Four Forces of Nature which are Gravity, Weak, Strong, and Electromagnetism. From these Oscillations, there are Energy Levels, and a Particle can be in many types of Energetic States. The Ground State is the lowest Energy State that a Particle can have, and that means that there is no state below the Ground State. A Particle can, however, get excited and arrive at Energy States above the Ground States. The greater the Energy the more the Particle vibrates with a smaller Wavelength and higher Frequency. The example in the last page shows how Energy States can be represented with Standing Waves. In order to map these Standing Waves for Particles, it would be necessary to visualize the Particle as Wave of Probability, where the higher the Amplitude of Oscillation the higher is the chance that the Particle is in that particular location, or having that particular Momentum at point in Time, or at an Energetic State. The Universe is mysterious for when it is not possible to be certain about what the Particles do, but the Cosmos still allows Scientists to make predictions based of Probabilities.

Love for all Things

Each one of us is comprised of an immense Grid of Atoms, Molecules, Cells, and flow of Substances. We breathe the air in and out, blood flows in our bodies, there is a constant chain of reactions between Molecules inside and between our Cells. When we share the same room with other people, we end up breathing the parts of the Air Molecules that they exhaled. The heat of people's bodies and their smell is absorbed by other people, which is proof of a great exchange of Particles between two or more people. When we see a person with our eyes, it means that Photons of Light emitted by that person's body went through our eyes' lens and is perceived by our brain igniting reactions between our Neurons. When we hear a person's voice, it is the disturbance made in that person's vocal cords that propagate through air and enters our ears with these vibrations which are then perceived by our brains. There are even Gravitational Forces between people. In truth nothing in the Universe is isolated and we are all strongly related to each other. There is Love in all of Creation. The immense heavens testifies that we are all connected and that is why we should better tolerate one another. The Ancient Greek Philosopher called Aristotle struggled to understand the concept of the one and the many. If all things are one, then why is it that from the one there are many? Science proved that all humans are related to each other and that we all descend from an African Women who lived thousands of years ago in Kenya. Also, Astrophysicists claim that from a Singularity smaller than a Proton all of the Universe was born. So, in fact one means many. Nothing in the Universe is in isolation. We are all one, and we are all individuals. The Constant Flux in the Universe between Particles generating Forces, Fields, and exchange of Energy is what binds all of the Cosmos into one single framework which is truly a work of art. Why then compete, or make war, or hate each other? In the end we all started in the same way, and we will continue bond together forever among Cosmic Dust in this great Dance of Atoms and

Particles. This is an ultimate realization that we all belong to a Cosmic Family. How can we make this family functioning better?

The Great Civilization:

When looking at an Ant Pile we don't normally see the ants fighting against each other but rather working together to fulfill a task which is to support their Kingdom and serve the Queen. We humans are more complex beings than ants but yet collaboration is often very difficult possibly because we are way more than ants, and there is a greater variety of ideas which may lead to conflicts. More complicated beings can make working together more difficult leading to competition and separation of a nation into tribes. The human ideas differ, and the construction of a human pile may not be done easily. In other words, ants can be better at fulfilling tasks since they are less evolved and are a simpler form of living beings. Being overly evolved like humanity adds difficulty to our interactions. Whether a Single World Civilization is possible or not is open for investigation, but history has proven that all World Civilizations were not able to last forever due to disputes, conflicts of ideas, and miscommunication. In that sense ants are better at building a civilization than humans despite our intelligence which is the main reason why we can't grow as a single nation. Being intelligent leads to disagreements.

Xhyn Universe Journey:

Suheep Hayver was enjoying his summer vacation in a ranch in the middle of a flat plane with broad green fields and few trees everywhere all around. It was located somewhere near the City of Roms and two Gnomes Yurgue and Zirvina made him a visit by bringing a Telescope and a book written by Ogo.

Hello Suheep Hayver! We have been very busy teaching Physics for several students. What we learn is that you can only be sure that you understand the subject when you are able to teach it. Sometimes you only truly learn the content when you start teaching it, since the questions that students ask can help you better prepare for the answers and explanations. This leads teachers to perform more research and better design lessons where students are exposed to diverse topics, situations, activities, and where learning is an opportunity not just for students but also for teachers.

In fact, learning is a way of living, since life is a great school. We participate in multiple scenarios to learn, and we breathe in and out while learning. There are multiple things around, and Reality is built from Perception. There are many Dimensions curled up at the Sub-Atomic world, and also at the Macro Scale at the size of Galaxies, Black Holes, and Universes. Maybe the world of the very small is just as complex as the world of the very big such as the size of our Multiverse. We then live in a Sea of Opportunities, and we navigate as we live with experiences, and nothing is settled, since all things are in constant motion following the flow of Time.

In this multiplicity, duality, singularity, and flow of fluids, the flow of Time, the flux of Energy, interactions between Particles and Beings, entanglement, and the motion through space is curled up forming the Space Time Reality that we live. We call this Reality a World, a Universe, a Dimension. Suheep Hayver you know all of this since you are a Knight of Ravens, one of the Guardians of Roms that fight monsters to guarantee the safety of the Great City and its people inside the Grid. Would you like to share a thought on our inspirations?

Yes, I am from the Grid, and I protect the Great City of Roms against all threats which makes me very close to Peter and Uflonix who are at the City Gates monitoring the flow of people in and out. I personally am fascinated by Science of anything but most especially about the mysteries of Space, the study of Astrophysics and also Chemistry which is the Study of Atoms, Mass, Energy, Matter, and the Chemical Reactions. These aspects of Nature permit life to flow, to move, to be part of our Existence and leads us to the love for all things.

Space is very big, and in fact it is everything that there is. Parts of Space are empty, and parts of it are filled with Matter. In fact, I have a list of pictures that Ogo took with his very small Monocular under high light pollution but that still allow us to have a good grasp for these immense heavens right above us. On the picture on the right, we have the great Ring Nebula. It is a Planetary Nebula formed after the death of a Star about the size of the Sun. It is around 2500 Light Years away from the Earth. First, we need to consider that our Galaxy has a Diameter of about 100,000 Light Years and a Light Year is the time that it takes light to move in that time. The Sun, for example, is 8 Light Minutes from the Earth and the other nearest Star called Alpha Centaury is 4 Light Years away.

The Ring Nebula is 2500 Light Years away, but given the size of our Galaxy, the Ring Nebula is a close neighbor. It really does look like a Ring, and it is beautiful since it appears unusual. Notice all of the other Stars around it. Space is dotted with Worlds and there are just a lot of places out there.

The Moon is 1.3 Light Seconds from the Earth, and it fascinates me. It is the closest Asteroid to Earth and is filled with craters and is a very silent desert with no wind, sound, or any sign of life anywhere except at the Poles where there is frozen water. Light travels at 300,000,000 meters per second, and the Moon is the closest Celestial object to us, but its light still takes over 1 second to reach us. Space is a Time Machine and everything we see in the sky is how these Celestial Objects were in the past. Gravity also travels at the Speed of Light which means that the Universe at the Past is what leads to Gravitational Forces in our present.

The Moon

The Hercules Globular Cluster also called Messier 13, contains hundreds of thousands of Stars very close together. This group of Stars has a Diameter of about 140 Light years and its Distance from Earth is of 23,000 Light Years. Globular Clusters can be found in any Galaxy and tend to be the oldest objects in the Universe and they act like satellites orbiting Galaxies. There is a lot of mystery with regard to these objects due to their old age, and how they were formed. Being inside a Globular Cluster would be like being exposed to a great amount of light, and many of these stars are as old as the Universe itself.

Hercules Globular Cluster

The Andromeda Galaxy is the closest Major Galaxy to our Milky Way with a Diameter of 150,000 Light Years and a Distance of 2.5 Million Light Years from Earth. In the picture we can only see the Center of the Galaxy that is very bright and a little bit of the very faint Nebulous Spiral that contains its Stars. This Galaxy is heading towards our Galaxy and in the future they will both merge into one. Galaxies contain 200 Billion of Stars and are like great Island Universes throughout Space. There are Trillion of Galaxies in the Universe with an enormous number of Worlds that we know nothing about. Are we alone in the Cosmos?

Andromeda Galaxy

Messier 8 also called the Lagoon Nebula is a large Interstellar Cloud that appears faint in the picture. It is still possible, however, to see a reddish color in it while it surrounds Star Cluster NGC 6530. It is about 4100 Light Years Away and is a great jewel in the Constellation Sagittarius. These Clouds are regions of Star Formation, and they are the remains of Stars that reached the end of their lives expelling gases into space. It is a great sight for Binoculars and from our viewpoint it is in the direction of the Center of our Galaxy. This is region of a lot of Galactic Clouds and a great density of Star Population.

Messier 8

The Orion Nebula is the best and most beautiful Nebula see from the Northern Hemisphere in the Constellation Orion. It is 1344 Light Years away from the Earth, and it is our closest region of Star Formation. The view is truly beautiful with several colors and many other Stars around it. It looks like a strange man with a dark cape looking down with bright lights coming from his bosom. The Stars around it seem like diamonds adorning the image that is one of the most notorious and most photographed regions in the night sky. Even under light pollution it is still truly gorgeous. To have more Nebulas like this we should hope for more star explosions in the future.

Orion Nebula

The Perseus Double Cluster is located in a densely populated region of the sky right along the band of the Milky Way. It is comprised of two clusters right next to each other and they are both about 7,500 Light Years away from us. These clusters are thought to be relatively young having been formed around 15 million years ago. An interesting fact is that these clusters are approaching the Earth appearing to show a blue shift in their spectrum. This is my favorite Star Cluster since it is so unique in the fact of being a double and the amount of stars that can be resolved while being very close together generates an awesome view and a great location to point the Telescope or Binoculars.

Perseus Double Cluster

The Dumbbell Nebula is a beautiful mostly green and blue cloud in space. It shows in the image at the right as a faint green smudge slightly left of the center of the picture. This Nebula is 1360 Light Years away from us. It is also a Planetary Nebula being it the remains of a star about the size of the Sun that reached the end of its life and releasing these gases in space. It is located inside the Summer Triangle comprise of the Stars Vega, Deneb, and Altair in the Constellation Vulpecula. What is so great about the view of this Nebula is how ghostly it looks and that is what makes it so beautiful to see. Nebulas are really my favorite locations in the sky.

Dumbbell Nebula

Wow Yurgue and Zirvina! Despite of the fact of the immense light pollution nowadays these pictures from Ogo's small monocular are able to show something in the sky. You showed me a total of 8 nice targets for star pictures and I enjoyed it. Stargazing is a fun hobby and makes we wonder about the true reason for these worlds to exist, or what the natural world is trying to tell us as these stars twinkle in the sky almost as if attempting to send us a message. Space is silent but we are still thrilled to try to listen to it, since nothing on my viewpoint is for no reason, and there must be an explanation for all the things we see up there. What do you think?

Definitely! The Stars are constantly sending us messages with their Gravitational Force, and also from the emission of Particles of Light that enter our eyes when we look at them. We can only see the Stars if their light emitted several years ago reaches your eyes. In other words, we are interacting with the Stars' Past when we look at them. Nothing can be seen or heard without interacting with what is emitting the Light and Sound. We can also convert the Light of Stars into Sound using a computer. Both Sound and Light are Waves and these are everywhere in this great Cosmic Sea of Particles.

Ogo also has a Theory of Cosmology that is very unique. Let us imagine a Universe in the shape of a Doughnut with arrows indicating expansion and contraction of the Cosmic Bubble through Space and Time.

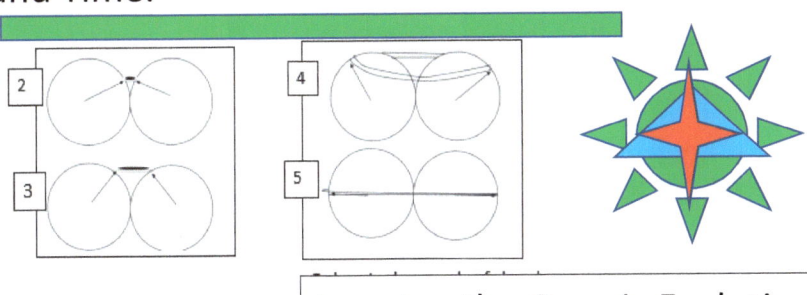

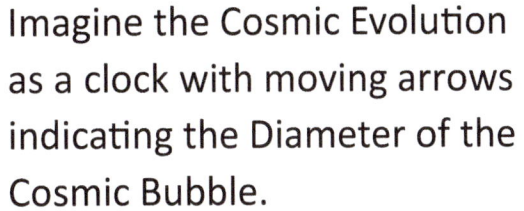

Imagine the Cosmic Evolution as a clock with moving arrows indicating the Diameter of the Cosmic Bubble.

ZIRVINA

Let us simulate a Universe that for us has a Life Expectancy of only 12.42 Seconds. Time is relative but we are outside of that Universe which for us lasts not for very long which is unlike for the beings who live in it.

For the following simulation, C is the Circumference of the wheels representing the Doughnut which in two Dimensions is shown as two Circles touching each other. θ is the Angle that the Arrows make from the Singularity at the Center where the two Circles meet. M is the Mass of the Universe, m is the Planck's Mass, and G is the Gravitational Constant. Let us pretend that our Universe is currently at Arrows Angle of $45°$ and r+ is the curved Distance from the tip of the Arrows to the Singularity at the Big Bang.

Now let us see what the Simulation will look like:

Simulation: A paper containing the following was left in Zeno's Mailbox

Let' say that a Small Universe is designed such that 10 m is the value of C and that we are currently located at $\theta = 45°$, and where GMm = 1 and m=1.

The integral for the Life Expectancy of the Universe becomes:

$$2\int_0^5 \sqrt{\frac{1}{\frac{1}{x}+\frac{1}{10-x}}}\, dx \qquad \text{where 2X5 = 10 =C}$$

The Integral equals: 12.42 s the life expectancy of our little universe: Very brief for us Outside of it.

The Current Velocity of the arrows at $\theta = 45°$ is with r+ equal to $\frac{\pi}{4}(r)$ where pi/4 = 45 degrees in Radians

r can be calculate from $C = 2\pi r = 10$

so r = $10/(2\pi)$ = 1.59 m

The Current Velocity is then:

$$V = \sqrt{\left(\frac{1}{x}+\frac{1}{10-x}\right)} = \text{with x being 12.5\% of a revolution}$$

since 45/360 = 0.125.

$$\sqrt{\left(\frac{1}{1.25}+\frac{1}{10-1.25}\right)} = 0.956 \text{ m/s current speed of arrows}$$

The Velocity of the arrows change with the size of Universe. When very small, arrows move fast, and when very large, arrows move slower.

It takes our Universe 12.42 s to move from Big Bang to Big Crunch.

The Velocity of Arrows Change with the Size of Universe:

Graphing : $\sqrt{(\frac{1}{x} + \frac{1}{10-x})}$ we get the following:

r+	Velocity of Moving Arrows
1	1.0540926
2	0.79056942
3	0.69006556
4	0.64549722
5	0.63245553

r+	Velocity of Moving Arrows
6	0.64549722
7	0.69006556
8	0.79056942
9	1.0540926

Except for the very beginning and end of a Cycle the Speed of Moving Arrows is between 1 m/s and 0.63 m/s with an Average of 0.805m/s.

Near the Beginning and the End the Graph:

$\sqrt{(\frac{1}{x} + \frac{1}{10-x})}$ shoots up. This explains Inflation in the Beginning of the Expanding Universe and Rapid Contraction at its End.

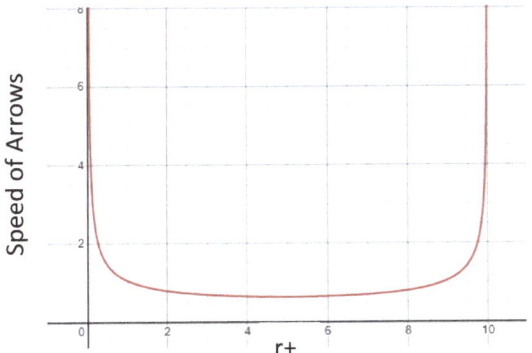

X-axis stands for r+ and Y-axis stands for Speed of Moving Arrows.

The Equation:

$$(1-1\sin(\theta/2))\ \Gamma\pi(\frac{R-R\cos(\theta)}{2})^2 = \text{Volume}$$

Has the Graph and Table in the Next Page:

Let us assume that $\Gamma = 0.01$ m the Thickness of our Cosmic Disc which is a mostly Flat Universe which is why a small number is chosen for it.

Big R = 2r

Our r = 1.59 m so R = 3.18 m

The equation is then:

$$10(1\text{-}1\sin(\theta/2))\ 0.01\pi(\frac{3.18-3.18cos(\theta)}{2})\text{^}2 = \text{Volume}$$

The 10 is placed in front of Equation to make the Graph appear more noticeable.

The Graph of that with changing θ of the Arrows is:

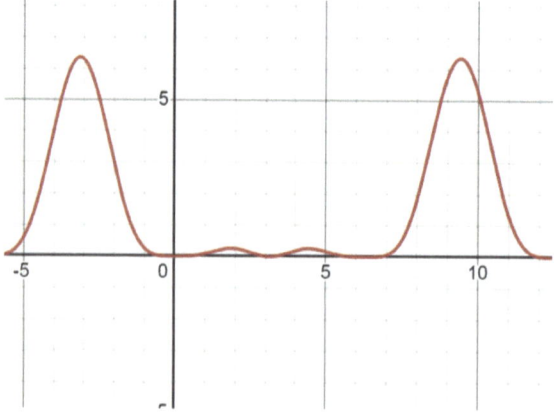

Only the Two Peaks in the Valley do we consider since the larger ones are just part of Quantum Fluctuations leading to no Universe.

The Table should only focus on Angles between 0 Degrees and 360 Degrees, before 0 Degrees and after 360 Degrees the graph shoots up too rapidly and these larger waves on the sides of the Valley cannot generate a Universe. Our Universe is in the Two Peaks in the Valley.

Angle	Volume in Meters Cubed	
$\frac{\pi}{4}$	0.042060137	Notice that although the Universe Expands and Contracts Twice within a Cycle, if we assume that we are at $\pi/4$ Radians we still see the Universe only expanding. That is exactly what we see in our Real Universe. After that, however, the Universe Contracts and becomes zero size at π Radians. An Infinitesimally Small Ring. After that, it Expands again becoming large at $6\pi/4$ Radians. It then Contracts once again reaching zero size at 2π Radians.
$\frac{2\pi}{4}$	0.23262342	
$\frac{3\pi}{4}$	0.17618419	
π	0	
$\frac{5\pi}{4}$	0.17618419	
$\frac{6\pi}{4}$	0.23262342	
$\frac{7\pi}{4}$	0.042060137	
2π	0	

As can be seen the Universe undergoes Expansion and Contraction Twice within a Cycle and that is why there are Two Peaks in the Valley. The contraction near half way through the Cycle is needed in order for the Cosmic Bubble to go around the Doughnut. It has to become an Infinitely Thin Ring in the middle of the Cycle in order to do that.

The Universe is a Thin Ring of Zero Size when the Angle is π Radians at the following stage:

In order for it to go around the Doughnut it gets smaller to allow it to make the turn around the corner.

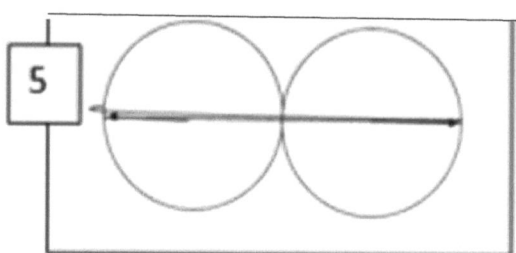

The Universe also has zero size at the very beginning and at its very end at zero and 2π Radians.

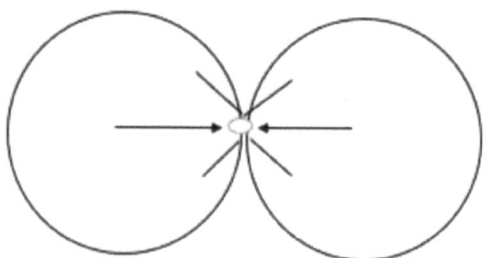

THE UNIVERSE IS A MACHINE THAT GOES INTO CYCLES WHICH CAN BE QUANTIFIED WITH EQUATIONS USING PHYSICS

Eureka! This simulation is for a small Universe that can fit in a Soccer Stadium. We played Simulation and this Universe lasts for only 12.41 seconds. The Bigger Universe which is our Real One lasts for Billions of Years. Time is relative however. For someone outside our Real Universe, the entire Cycle may indeed last for only 12.41 seconds just like in our Simulation.

Life is beautiful and all things have a meaning that can be explained through Mathematics. Feelings are caused by Perception which is nothing but Hormones and Fluids within our Biological Bodies. These Glands can be thoroughly understood with Numbers, Quantities, and burst of Energy at given amounts. The Universe is curled up on itself and it Creates and Destroys itself through Cycles. Existence is a Cycle, and all things vibrate at given Frequencies. What makes you different than me is a Frequency. Each Soul is an Atom, a Particle of Light, and we use our bodies for a temporary existence. From dust everything came into being and to dust all shall return. This is the flow of Time and the Cycle of Life, the most beautiful story ever told is our Life Stories.

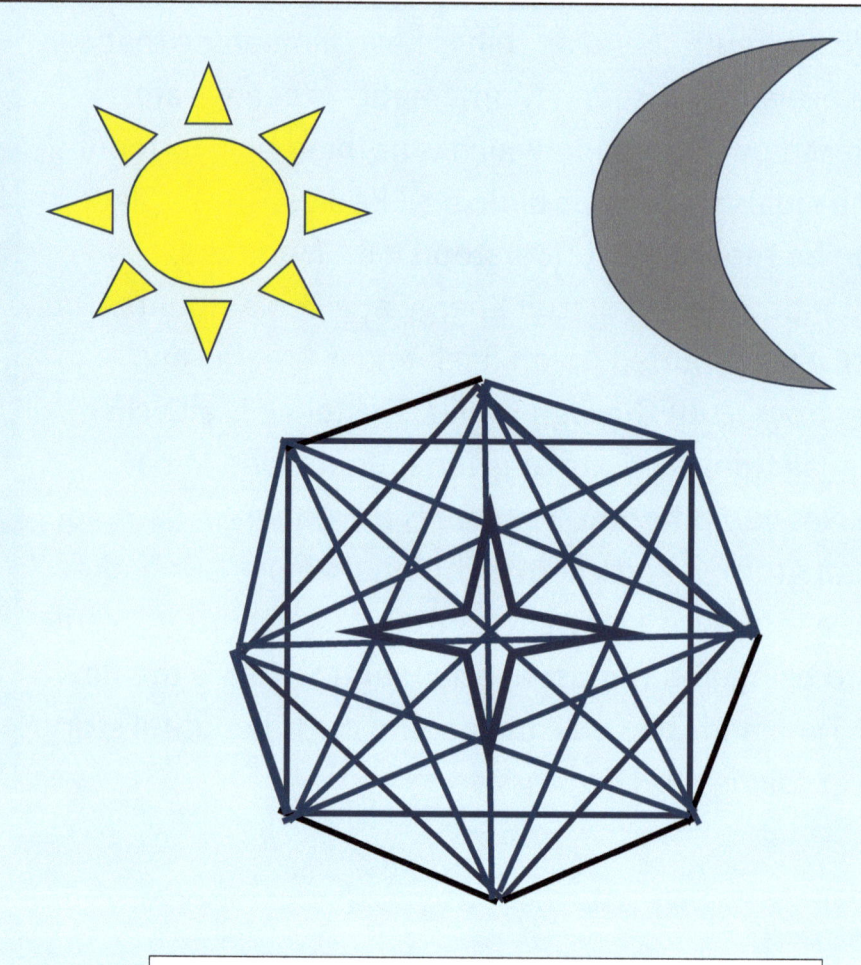

All Things Curl on itself and every Particle is Entangled to every other Particle.

The Universe is ONE.

<u>About the Author</u>

I, Diogo Franklin de Souza, was born in the city of Rio de Janeiro, Brazil in August 20, 1986. I moved to Dallas, Texas when I was 11 years old. I write stories since I was 9 years old. My books tend to contain short summaries of the most important things I find about life, morality, religion, philosophy, and science. Like I say, everything is part of a whole system, and this is also for everything I do and write. I always wanted to have all the most important knowledge in only a few short books. That is why I write, and that is my inspiration for short summaries. I hope this book brings some inspiration also for the readers, because that really is the purpose of my work. Read it and take from it, pieces of gold for you that can be useful in your life. Enjoy….